Bullfrog Books

Farm Animals

Turkeys on the Farm

by Bizzy Harris

Ideas for Parents and Teachers

Bullfrog Books let children practice reading informational text at the earliest reading levels. Repetition, familiar words, and photo labels support early readers.

Before Reading
- Discuss the cover photo. What does it tell them?
- Look at the picture glossary together. Read and discuss the words.

Read the Book
- "Walk" through the book and look at the photos. Let the child ask questions. Point out the photo labels.
- Read the book to the child, or have him or her read independently.

After Reading
- Prompt the child to think more. Ask: What did you know about turkeys before reading this book? What more would you like to learn about them?

Bullfrog Books are published by Jump!
5357 Penn Avenue South
Minneapolis, MN 55419
www.jumplibrary.com

Copyright © 2021 Jump! International copyright reserved in all countries. No part of this book may be reproduced in any form without written permission from the publisher.

Library of Congress Cataloging-in-Publication Data

Names: Harris, Bizzy, author.
Title: Turkeys on the farm / by Bizzy Harris.
Description: Minneapolis, MN: Jump!, Inc., [2021]
Series: Farm animals | Includes index.
Audience: Ages 5–8 | Audience: Grades K–1
Identifiers: LCCN 2020021761 (print)
LCCN 2020021762 (ebook)
ISBN 9781645277156 (hardcover)
ISBN 9781645277163 (ebook)
Subjects: LCSH: Turkeys—Juvenile literature.
Livestock—Juvenile literature.
Classification: LCC SF507 .H37 2021 (print)
LCC SF507 (ebook) | DDC 636.5/92—dc23
LC record available at https://lccn.loc.gov/2020021761
LC ebook record available at https://lccn.loc.gov/2020021762

Editor: Eliza Leahy
Designer: Molly Ballanger

Photo Credits: khak/Shutterstock, cover; photomaster/Shutterstock, 1, 16, 22, 23tm; Lepas/Shutterstock, 3; Green Mountain Montenegro/Shutterstock, 4 (background); Willeecole/Dreamstime, 4 (foreground), 23bm; Denys Prokofyev/Dreamstime, 5; Steven Kazlowski/SuperStock, 6–7; NataliaVo/Shutterstock, 8–9; Amy McNabb/Shutterstock, 10–11; Redteniks/Dreamstime, 12, 23tr; Charlotte Bleijenberg/Shutterstock, 13, 23tl, 23br; Pornsawan Baipakdee/Shutterstock, 14–15; Axente Vlad/Shutterstock, 17; Blanscape/Shutterstock, 18–19, 23bl; Design Pics/SuperStock, 20–21; Tsekhmister/Dreamstime, 24.

Printed in the United States of America at Corporate Graphics in North Mankato, Minnesota.

Table of Contents

Beaks and Bugs 4

Parts of a Turkey 22

Picture Glossary 23

Index 24

To Learn More 24

Beaks and Bugs

What will hatch from this egg?

A baby turkey!
We call it a poult.

... poult

The hen is the mom.
Her poults stay close.

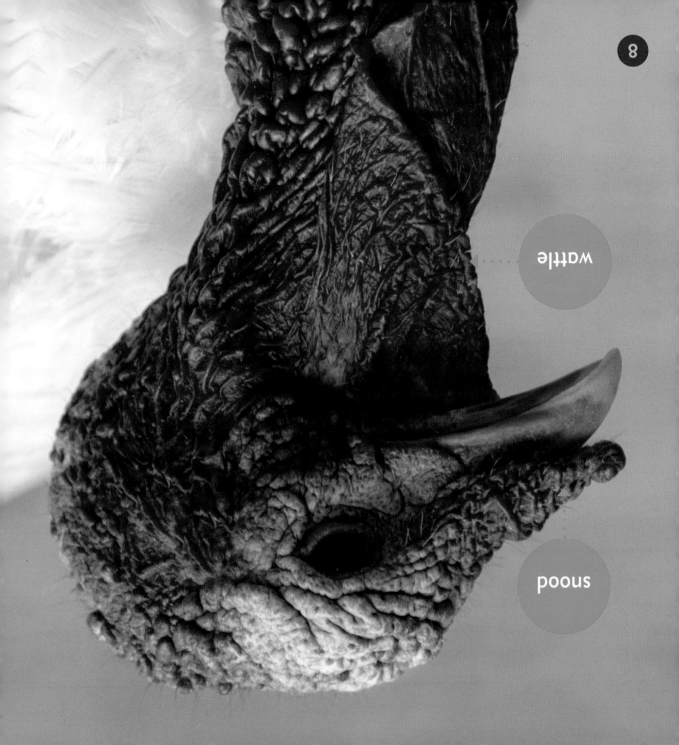

Each turkey has a wattle.

What is that on its beak?

A snood!

Neat!

Toms are males.
They have beards.
They can get long!

beard

Toms gobble.

tail

This one struts.
He fans his tail.
Wow!

Some turkeys are white.
Others are brown.
This one has blue feathers.
Cool!

They flap their wings.
But they cannot fly far.
They are too big!

They run fast.
Their strong legs help.
Zoom!

Turkeys graze.
They also eat bugs.

They need a rest!
They sit up high.

Parts of a Turkey

Take a look at the parts of a tom!

22

Picture Glossary

fans
Spreads out like a fan.

flap
To move up and down.

gobble
To make a natural noise.

graze
To feed on grass that grows in a field.

hatch
To break out of a shell.

struts
Walks with a swagger or in a proud way.

Index

bugs 18
eat 18
egg 4
fly 16
gobble 12
graze 18
hatch 4
hen 6
poult 5, 6
run 17
struts 13
toms 10, 12

To Learn More

Finding more information is as easy as 1, 2, 3.

① Go to www.factsurfer.com
② Enter "turkeysonthefarm" into the search box.
③ Choose your book to see a list of websites.